DAY AND NIGHT IN THE

Rain Forest

by Ellen Labrecque

PEBBLE

a capstone imprint

Published by Pebble, an imprint of Capstone.
1710 Roe Crest Drive, North Mankato, Minnesota 56003
capstonepub.com

Library of Congress Cataloging-in-Publication Data
Names: Labrecque, Ellen, author.
Title: Day and night in the rain forest / by Ellen Labrecque.
Description: North Mankato, Minnesota : Pebble, [2022] | Series: Habitat days and nights | Includes bibliographical references and index. | Audience: Ages 5-8 | Audience: Grades K-1 |
Summary: "Spend a day and night in the rain forest! Learn about this lush habitat through the diverse animals that call it home. Start the morning suspended high in the canopy with a colorful toucan. Curl around a branch and bask in afternoon sun with an emerald boa. At sunset, pace the forest floor for prey alongside a sleek jaguar. After dark, spy a nocturnal sloth slowly wake after a full day of slumber. What will tomorrow bring in the rain forest?"-- Provided by publisher.
Identifiers: LCCN 2021041496 (print) | LCCN 2021041497 (ebook) |
  ISBN 9781663976918 (hardcover) | ISBN 9781666327793 (paperback) |
  ISBN 9781666327809 (pdf) | ISBN 9781666327823 (kindle edition)
Subjects: LCSH: Rain forest animals--Behavior--Juvenile literature. | Habitat (Ecology)--Juvenile literature.
Classification: LCC QL112 .L3285 2022 (print) | LCC QL112 (ebook) | DDC 591.734--dc23
LC record available at https://lccn.loc.gov/2021041496
LC ebook record available at https://lccn.loc.gov/2021041497

Image Credits
iStockphoto: MmeEmil, Cover (scarlet macaw), 1, slowmotiongli, 17; Mighty Media, Inc.: 20, 21; Shutterstock: buteo, 18, DanielSnake, 13, Dennis Stogsdill, 7, Dirk Ercken, 14, Fotos593, 5, Jamen Percy, 15, Kevin Wells Photography, 11, kingma photos, 9, Leonardo Mercon, 16, MarcusVDT, 8, Ondrej Prosicky, 19, Photo Spirit, 6, pyzata, Cover (rain forest), 1, Ryan M. Bolton, 10

Editorial Credits
Jessica Rusick, editor, media researcher; Kelly Doudna, designer, production specialist

All internet sites appearing in back matter were available and accurate when this book was sent to press.

Table of Contents

Words in **bold** are in the glossary.

What Is a Rain Forest?

Rain forests are wet forest **habitats**. They have many tall trees. There are rain forests around the world. Some are in Central and South America.

Many animals live in rain forests. Some come out during the day. Others come out at night.

A rain forest in South America

Morning

The sun rises over the rain forest. Spider monkeys swing through the trees. They grab branches with their long arms and tails. The monkeys search for fruit to eat.

Spider monkey

Toucan

A colorful toucan makes loud croaking sounds. It is talking to other toucans! The sound **echoes** across the rain forest.

Noon

A green iguana rests in the sun. Iguanas are cold-blooded. They need the sun to warm their bodies.

The iguana sits high in a tree. Suddenly, an eagle swoops down. It tries to catch the iguana! The iguana whips its tail. It hits the eagle. The bird flies away.

Eagle

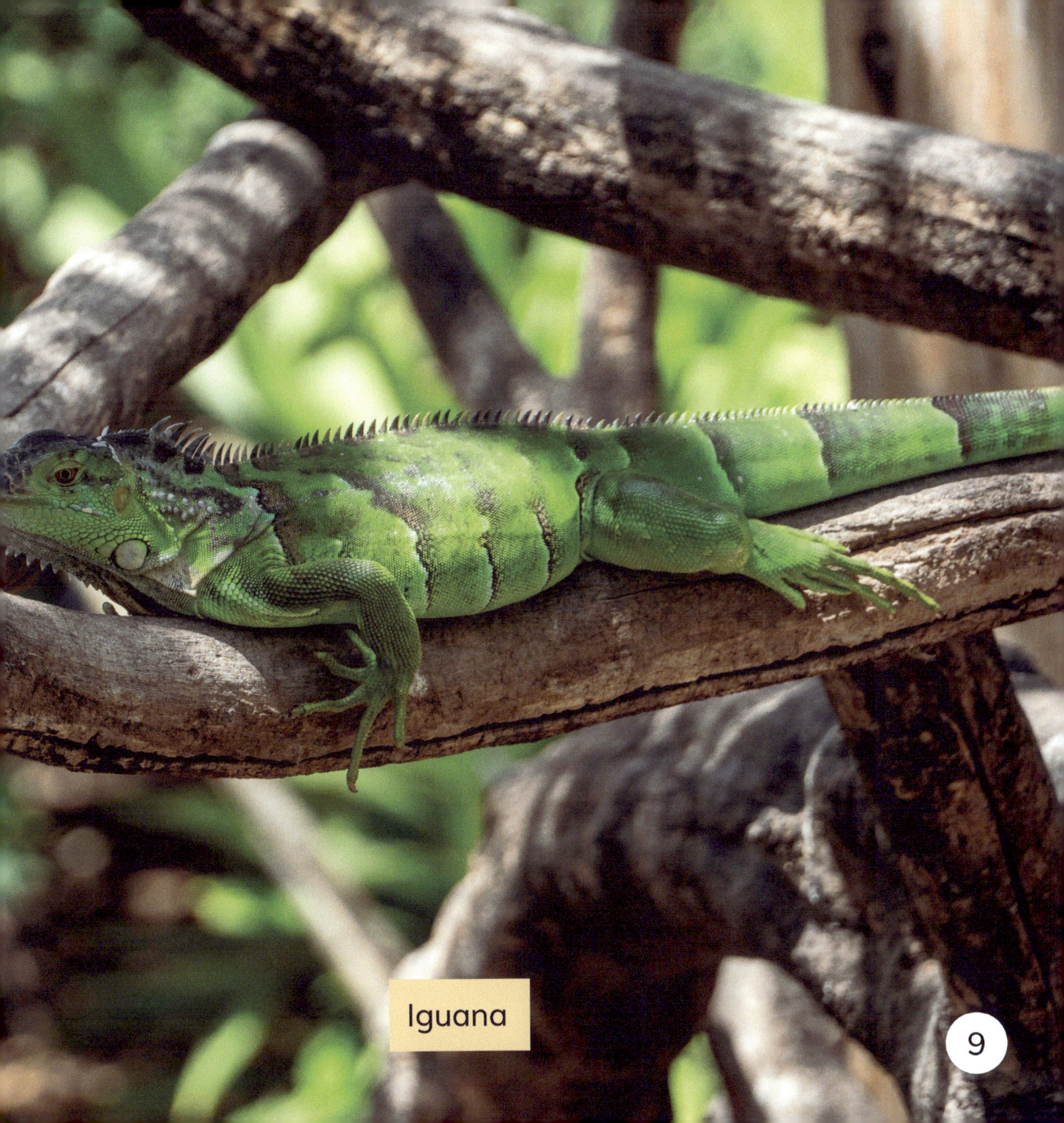

Iguana

Late Afternoon

A bridled forest gecko hunts for bugs. It stays in the shade to keep cool. The gecko sees an ant. It flicks out its sticky tongue to catch the ant!

Bridled forest gecko

Red-eyed tree frog

A red-eyed tree frog sleeps on a leaf. The frog sleeps during the day. Its green skin blends in with the leaf. This helps the frog hide from **predators**.

Evening

The sun sets. Many creatures return to their nests and **burrows**. Other animals wake up.

A tarantula comes out of its burrow. It is the size of a dinner plate! The tarantula hunts for bugs. It grabs an ant with its front legs. Then it crushes the ant in its mouth.

Tarantula

13

Night

An emerald tree boa is draped over a tree branch. It slept like this during the day. At night, the snake lowers its head and neck. It bites **prey** that comes near!

Emerald tree boa

Jaguar

A jaguar silently hunts. It can see well in darkness. The jaguar follows its prey from the shadows. Then, the big cat **pounces**! It can kill prey with one bite.

Late Night

An armadillo crawls on the forest floor. Its hard shell protects it from predators. The armadillo digs a hole with sharp claws. It is looking for bugs to eat.

A sloth wakes up after sleeping all day. It lazily eats a wet leaf. Sloths are the slowest **mammals** on Earth! The sloth mostly stays in its tree.

Armadillo

Sloth

Dawn

The sun rises. A capybara eats grass by a river. Suddenly, it hears a predator. The capybara hides in the river. It can stay underwater for five minutes!

Capybara

Scarlet macaws

Two scarlet macaws **perch** on a branch. The birds clean each other's feathers with their beaks. Chirps and growls fill the air. Another day in the rain forest has begun.

Tree Frog Activity

What You Need:

- paper plate
- green marker or paint
- cup
- pencil
- red construction paper
- scissors
- black marker
- glue

What You Do:

1. Color or paint both sides of a paper plate green.

2. Fold the paper plate in half.

3. Use a cup to trace two circles on red paper. Cut out the circles.

4. Draw two large black diamonds inside the circles.

5. Glue the circles to the flat edge of the folded paper plate. These are the frog's eyes.

6. Cut a long strip of red paper. Glue one end inside the paper plate so it sticks out. This is the frog's tongue. Pretend to catch bugs with your paper plate frog!

Glossary

burrow (BUHR-oh)—a tunnel or hole in the ground made or used by an animal

echo (EK-oh)—to repeat over and over again

habitat (HAB-uh-tat)—the natural place and conditions in which a plant or animal lives

mammal (MAM-uhl)—a warm-blooded animal that breathes air; mammals have hair or fur; female mammals feed milk to their young

perch (PURCH)—to stand on the edge of something

pounce (POUNSS)—to jump on something suddenly and grab it

predator (PRED-uh-tur)—an animal that hunts other animals for food

prey (PRAY)—an animal hunted by another animal for food

Read More

Barnham, Kay. *Incredible Rain Forests*. New York: Crabtree Publishing Company, 2021.

Delano, Marfé Ferguson. *Rain Forests*. Washington, D.C.: National Geographic Kids, 2017.

Messner, Kate. *Over and Under the Rainforest*. San Francisco: Chronicle Books, 2020.

Internet Sites

DK Find Out!—Amazon Rain Forest
dkfindout.com/us/animals-and-nature/habitats-and
-ecosystems/amazon-rain-forest/

National Geographic Kids—Rainforest Habitat
kids.nationalgeographic.com/nature/habitats/article
/rain-forest

Rainforest Alliance—Kids' Games & Activities
rainforest-alliance.org/kids

Index

About the Author

 Ellen Labrecque is the author of more than 100 nonfiction children's books. She lives in Bucks County, Pennsylvania, with her husband and two kids. She has the best writing partner in the world—her dog, Oscar. An avid reader and runner, Ellen is a morning person. On most days, she is up before the sun.